Interactive
Mathematics Program®

INTEGRATED HIGH SCHOOL MATHEMATICS

The World of Functions

FIRST EDITION AUTHORS:
Dan Fendel, Diane Resek, Lynne Alper, and Sherry Fraser

CONTRIBUTORS TO THE SECOND EDITION:
Sherry Fraser, IMP for the 21st Century
Jean Klanica, IMP for the 21st Century
Brian Lawler, California State University San Marcos
Eric Robinson, Ithaca College, NY
Lew Romagnano, Metropolitan State College of Denver, CO
Rick Marks, Sonoma State University, CA
Dan Brutlag, Meaningful Mathematics
Alan Olds, Colorado Writing Project
Mike Bryant, Santa Maria High School, CA
Jeri P. Philbrick, Oxnard High School, CA
Lori Green, Lincoln High School, CA
Matt Bremer, Berkeley High School, CA
Margaret DeArmond, Kern High School District, CA

Key Curriculum Press

Second Edition   I M P

This material is based upon work supported by the National Science Foundation under award numbers ESI-9255262, ESI-0137805, and ESI-0627821. Any opinions, findings, and conclusions or recommendations expressed in this publication are those of the authors and do not necessarily reflect the views of the National Science Foundation.

Key Curriculum Press
1150 65th Street
Emeryville, California 94608
email: editorial@keypress.com
www.keypress.com
10 9 8 7 6 5 4 3 2 1   15 14 13 12 11
ISBN 978-1-60440-145-5
Printed in the United States of America

Project Editors
Mali Apple, Josephine Noah

Project Administrator
Emily Reed

Professional Reviewers
Rick Marks, Sonoma State University, CA
D. Michael Bryant, Santa Maria High School, CA, retired

Accuracy Checker
Carrie Gongaware

First Edition Teacher Reviewers
Kathy Anderson, Aptos High School, CA
Dan H. Brutlag, Tamalpais High School, CA
Robert E. Callis, Hueneme High School, CA
Susan Schreibman Ford, Delhi High School, CA
Mary L. Hogan, Arlington High School, MA
Jane M. Kostik, Patrick Henry High School, MN
Brian Lawler, California State University San Marcos, CA
Brent McClain, Vernonia School District, OR
Michelle Novotny, Eaglecrest High School, CO
Barbara Schallau, East Side Union High School District, CA
James Short, Oxnard Union High School District, CA
Kathleen H. Spivack, Wilbur Cross High School, CT
Linda Steiner, Orange Glen High School, CA
Marsha Vihon, Corliss High School, IL
Edward F. Wolff, Arcadia University, PA

First Edition Multicultural Reviewers
Genevieve Lau, Ph.D., Skyline College, CA
Luís Ortiz-Franco, Ph.D., Chapman University, CA
Marilyn Strutchens, Ph.D., Auburn University, AL

Copyeditor
Brandy Vickers

Interior Designer
Marilyn Perry

Production Editor
Andrew Jones

Production Director
Christine Osborne

Editorial Production Supervisor
Kristin Ferraioli

Compositors
Kristin Ferraioli, Maya Melenchuk

Art Editor/Photo Researcher
Maya Melenchuk

Technical Artists
Lineworks, Inc., Maya Melenchuk, Kristin Ferraioli

Illustrator
Juan Alvarez, Alan Dubinsky, Tom Fowler, Nikki Middendorf, Briana Miller, Evangelia Philippidis, Paul Rodgers, Sara Swan, Martha Weston, April Goodman Willy, Amy Young

Cover Designer
Jenny Herce

Printer
Lightning Source, Inc.

Executive Editor
Josephine Noah

Publisher
Steven Rasmussen

Contents

The World of Functions—Families of Functions and the Algebra of Functions

The World of Functions

Families of Functions and the Algebra of Functions

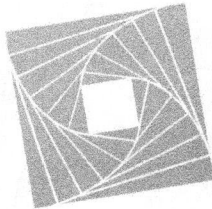

The World of Functions—Families of Functions and the Algebra of Functions

The What and Why of Functions

This unit gives you the opportunity to look back over your past work in mathematics and examine the roles played by functions. You will pull together ideas about functions that you studied previously, and you will learn new ways of thinking about functions.

You will begin by recalling what you know about functions and then trying to make sense of some data on stopping distances.

Giancarlo Beroldo looks back at the role functions played in a Year 3 unit.

Brake!

A good driver should know how far the car will go after he or she applies the brakes. The table gives a certain car's stopping distance as a function of the car's speed at the time the driver hits the brakes. Here, "stopping distance" means the distance the car travels from the moment the driver actually begins applying the brakes until the car stops.

Speed (in miles per hour)	Stopping distance (to the nearest tenth of a foot)
20	22.2
25	34.7
30	50.0
35	68.0
40	88.8
45	112.4
50	138.8

Based on this table, figure out as much as you can about how stopping distance is affected by speed. Your work should include each of these elements:

- A graph of the data set, with appropriate scales and labels for the axes

- A verbal description of any patterns you see in the data set

- A prediction of what the stopping distance for this car would be for a speed of 70 miles per hour, with an explanation of how you made your prediction

One Mile at a Time

You may have noticed that many highways have small signs along the side of the road that list your mileage from some place behind you. This POW concerns a situation involving these mileposts.

The Situation

You are driving down a highway and notice a milepost showing a distance that is a two-digit number. Exactly one hour later, you notice a milepost that shows the same two digits, but in the opposite order.

Then, exactly one hour after passing the second milepost, you see a third one that shows a three-digit number. The middle digit is 0. The other two digits are the same as those on the first milepost and are in the same order as on that first milepost.

Your Task

The basic question for the situation is, "How fast have you been traveling?" (Assume you have been traveling the same speed throughout your trip.) This question has a simple numeric answer that you may be able to find by guess-and-check.

Your task is to go beyond finding the answer, by expressing the situation in algebraic terms and proving, using algebra, that your solution is correct. Recall that a number like 386 can be written as $3 \cdot 100 + 8 \cdot 10 + 6$. Apply this idea to express an arbitrary two- or three-digit number in terms of its digits.

Write-up

1. *Process*

2. *Solution:* Answer the question and give a complete explanation of how you can be sure your answer is correct. Your explanation is the most important part of this POW.

3. *Self-assessment*

Story Sketches

This activity presents several situations, each described in fairly general terms. In each case, you will draw a "story sketch" of what the graph might look like.

If no particular numbers are given, make the situations more specific by making specific numeric assumptions so you can use scales on the axes of your graphs. Explain any assumptions you make in developing each graph sketch.

1. Students are putting on a show. Sketch a graph showing their overall profit as a function of the number of tickets they sell.

2. You ride down in the elevator from the top floor of a 20-story building without making a stop. Sketch a graph showing your height off the ground as a function of the time you have spent on the elevator since it began moving.

3. You invested $1000 in a savings account that pays four percent interest per year. Sketch a graph showing the amount of money in the account as a function of the time elapsed since you made the investment.

4. A colony of bacteria is beginning to grow in someone's lung. Starting with a specific number of bacteria, chart their growth in a table of values over time, and then make a graph.

5. Make up a situation of your own about some function, and draw that function's graph.

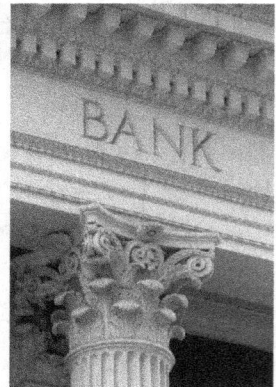

Story Sketches II

As in *Story Sketches,* your task here is to develop a sketch that shows the general shape of a graph for the given situation. Again, explain the assumptions you make in developing each sketch. Be sure to provide scales for each axis.

1. Someone is riding on a Ferris wheel. Sketch a graph of that person's height above the ground over time.

2. The amount of sunlight in a day changes throughout the year. Sketch the number of hours of sunlight in a day as a function of the time of year.

3. A certain radioactive material loses half the carbon it contains every 20 years. Sketch a graph showing the amount of carbon left as a function of time.

4. A ball is dropped off the roof of a building. It hits the ground, bounces up, goes back down, bounces up again, and so on.

   Each time it bounces, it loses some of its height. Make a table showing how high the ball goes on each bounce. (*Note:* This is different from a table showing the height of the ball over time.)

5. Make up a situation whose graph has the same general shape as one of the graphs for Questions 1 to 4. Make your situation as different as possible from that situation.

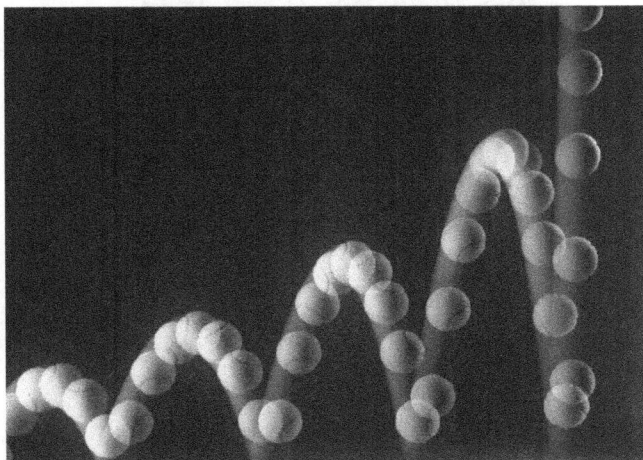

What Good Are Functions?

You have met many functions in connection with the solution of unit problems. In this activity, you will recall some ways in which functions were helpful to you.

Select a unit you studied previously and a specific function from that unit. Then do these three things:

- Describe the problem context in which the function was used, and explain what the input and output for the function represent in terms of the problem context.

- Describe how the function was helpful to you in solving the central unit problem or some other problem in the unit.

- If possible, tell which family the function is from. (You don't need to give the exact equation for the function.)

If you have time, do the same things for some other units.

More Families

As you have seen, functions from the same family have several things in common.

- Their graphs have a roughly similar shape.
- Their equations can be put into similar forms.
- The real-world situations that relate to them have common properties.

The two problems in this activity continue your work with families of functions.

1. An object is dropped from a cliff. As the object falls, its height off the ground below is a function of how long it has been falling.

   a. Sketch a graph of the function, listing any assumptions you make.

   b. Name the function family involved.

   c. Describe the general algebraic form for functions in this family.

2. A sporting-goods company packages the balls it produces in cube-shaped boxes. The volume of a cube is a function of the length of its sides.

   a. Sketch a graph of the function, listing any assumptions you make.

   b. Name the function family involved.

   c. Describe the general algebraic form for functions in this family.

Tables

You have seen that one way of representing a function is with a table. Your next step is to figure out how to tell which function family certain tables belong to. In the process, you will prove some of the properties of tables that you discover for these families.

Simon Bishop and Erik Santillan use their calculators to look for patterns in tables in order to identify different families of functions.

Linear Tables

One way to look at a function is through a table of values. In this activity, you will look specifically at tables of **linear functions**—functions described by an equation of the form $y = ax + b$.

In particular, you will explore the relationship between patterns in the tables of linear functions and other aspects of these functions, especially their algebraic form.

1. Start with the specific function $f(x) = 4x + 7$.
   a. Create an In-Out table for f using equally spaced inputs.
   b. Look for a pattern in the *Out* values.

2. Now consider other specific linear functions.
   a. Examine the table of each function using equally spaced inputs, and look for patterns. Use a variety of "equal spacings."
   b. Formulate a general statement of a pattern that holds true in the tables of all linear functions.

3. Use the algebraic form $y = ax + b$ for linear functions to prove your results from Question 2.

4. Explain why the patterns you found in Question 2 make sense in terms of the graphs of linear functions.

5. Create a real-world situation that is described by a linear function. Explain why the pattern from Question 2b makes sense in terms of that situation.

Story Sketches III

In this activity, you will sketch graphs or make tables for more real-world situations. As before, explain any assumptions you make about the situations in order to create your graphs or table.

You may also choose to assign numeric values to certain parts of the problems if that will clarify your work. Include scales on the axes of your graphs.

1. You are driving due north on the highway. Ahead, but off to the east, you see a tall tower. Sketch a graph or make a table showing your distance from the tower as a function of time as you continue north.

2. You're preparing a pie for a family dinner. You turn on the oven and set it at 325° Fahrenheit. Sketch a graph or make a table of the temperature in the oven as a function of time as the day goes on.

3. Each morning, you write down the time at which the sun rises. Make a table or sketch a graph showing how that time varies over the course of the year.

Quadratic Tables

You've seen that In-Out tables for linear functions all have a special property and that this property can be proved in terms of the algebraic form of linear functions.

Now you will look for a similar property in the tables for functions in the quadratic family. In later activities, you'll examine whether this new property can be proved in terms of the algebraic form of functions in that family.

1. Start by choosing a specific quadratic function. Create an In-Out table for that function using equally spaced inputs, and then look for patterns. Consider both positive and negative inputs.

2. Continue with other functions in that family, looking for some type of pattern that holds true for the tables of all quadratic functions. Consider tables in which the difference between inputs is a value other than 1. Also consider quadratic functions in which the coefficient of x^2 is a value other than 1.

3. Examine how the pattern you found in Questions 1 and 2 varies from one quadratic function to another. State your observations in as general a way as you can.

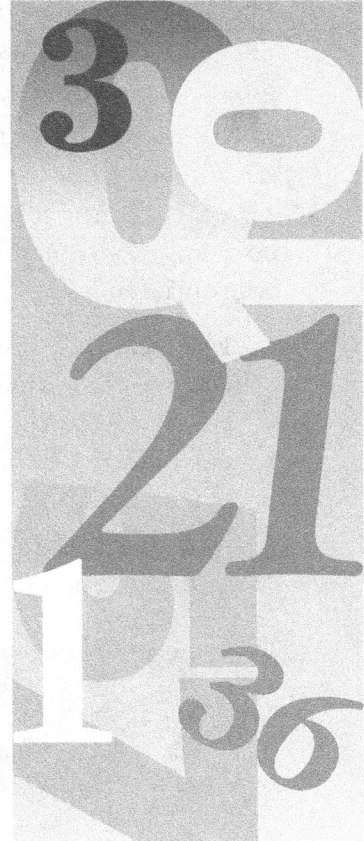

Back to Basics

The family of linear functions is a basic part of the world of functions. This activity presents four real-world situations in which someone might consider using a linear model.

In Questions 1 to 4, use a table, a graph, or an algebraic expression to explain how to solve the problem using the assumption of linearity. If a problem is impossible to solve, explain why, and describe how it can be changed so it can be solved. In Question 5, you will consider the appropriateness of the linear model in each case.

1. Ms. Jackson is buying scientific calculators for her math class. She tells the principal that a classroom set of 30 calculators would cost $388.50. The principal wants to know the cost of purchasing calculators for the entire school. How much would it cost to buy 300 calculators?

2. A hiker is climbing steadily up a mountain. If he reaches an altitude of 3000 feet at 3 p.m., what will his altitude be at 6 p.m.?

continued ▶

3. Jacob and Kenisha each rent a car for the day from the Rent-the-Best car rental agency. Rent-the-Best charges a daily minimum rate plus a per-mile charge. Jacob drives 200 miles and pays $41. Kenisha drives 250 miles and pays $45.

   If your budget allows $50 for car rental for the day, how many miles can you afford to drive if you use Rent-the-Best?

4. Chloe is spending $75 a day on her vacation. Day 6 finds her lying on a Caribbean beach with $784 left to spend. Her plans have her returning home on Day 10. How much money will she have left then?

5. For each of the situations in Questions 1 to 4, answer these two questions.

   a. What clues tell you that the problem, as written, represents a linear situation?

   b. Does the problem describe a situation that would be linear in real life?

Quadratic Tables by Algebra

In this activity, you will prove algebraically that a specific quadratic function has constant second differences, at least for the case in which the inputs differ by 1. Question 1 examines the function numerically, and Question 2 represents the general proof.

1. Consider the function $f(x) = x^2 + 2x + 3$.

   a. Complete the In-Out table for this function.

x	$f(x)$
7	
8	
9	
10	

   b. Find the differences between successive outputs in the table, and then find the second differences.

2. Consider the same function as in Question 1. But now, suppose the first input in the table is represented by the variable w. Subsequent inputs are expressed in terms of w, as shown in the next table.

   a. Complete the table.

   b. Find the differences between successive outputs in the table, and then find the second differences.

x	$f(x)$
w	
$w + 1$	
$w + 2$	
$w + 3$	

   c. Describe the connection between your answers to Question 2b and your answers to Question 1b.

   d. Explain why your results prove that the function f has constant second differences.

A General Quadratic

In *Quadratic Tables by Algebra,* you looked at part of an In-Out table for the function f defined by the specific equation $f(x) = x^2 + 2x + 3$. Now consider a general quadratic function, given by the equation $g(x) = ax^2 + bx + c$ (where $a \neq 0$).

1. First consider a table using the same numeric inputs you used in *Quadratic Tables by Algebra,* as shown here, but using the general quadratic function g.

   a. Complete the In-Out table.

x	$g(x)$
7	
8	
9	
10	

   b. Find the differences between successive outputs in the table, and then find the second differences.

2. Now use w, $w + 1$, and so on as the inputs, as shown in the next table.

x	$f(x)$
w	
$w + 1$	
$w + 2$	
$w + 3$	

   a. Complete the In-Out table (still using the general quadratic function g).

   b. Find the differences between successive outputs in the table, and then find the second differences.

3. Describe in specific terms how your answers to Questions 1 and 2 relate to your answers to Questions 1 and 2 of *Quadratic Tables by Algebra.*

Exponential Tables

You have discovered that In-Out tables for linear and quadratic functions have special properties. You have also learned that those properties can be proved in terms of the algebraic form of the functions in those families.

You will now look for properties of tables for the **exponential function** family and explain those properties in terms of the algebraic form of functions in that family.

For linear and quadratic functions, you found patterns by subtracting consecutive outputs. In this activity, you will find two patterns: one by subtracting consecutive outputs and another by dividing consecutive outputs.

1. First look for patterns by *dividing* consecutive outputs.

    a. Start by choosing a specific exponential function. Create an In-Out table for that function using equally spaced inputs. Look for patterns in the *ratios* of consecutive outputs.

    b. Continue with other exponential functions. Look for some type of pattern that holds true in the tables of all exponential functions.

2. Explain your results in terms of the type of equation used to express exponential functions. Give as general a proof of your results as you can.

continued ▶

3. Now look for a second pattern in the table by *subtracting* consecutive outputs.

 a. Start by choosing a specific exponential function. Create an In-Out table for that function, using equally spaced inputs. Then look for patterns in the *differences* of consecutive outputs.

 b. Continue with other exponential functions. Look for some type of pattern that holds true in the tables of *all* exponential functions.

A Cubic Pattern

You have looked at patterns for the tables of functions in both the family of linear functions and the family of quadratic functions. Differences and second differences for the outputs of these tables played important roles in those patterns.

Now consider the family of **cubic functions.** This is the set of functions defined by equations of the form $f(x) = ax^3 + bx^2 + cx + d$ (with $a \neq 0$).

Specifically, consider the simplest function in this family, which is given by the equation $f(x) = x^3$.

1. Make an In-Out table for this function (using equally spaced inputs).

2. Find a pattern in your table and clearly state it.

3. Use algebra to prove that pattern for the case in which the inputs differ by 1.

Betty, Al, and Carlos are playing games with the three spinners shown here. The game is played by having each participant spin his or her spinner. The player with the highest number is the winner. Because the numbers on the spinners are all different, there are never any ties.

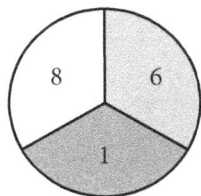

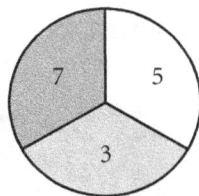

Betty's spinner	Al's spinner	Carlos's spinner

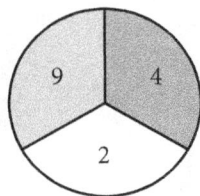

Each spinner is divided into three equal parts, so each numeric result on a given spinner is equally likely. (The size of the winning number is not important. For instance, if Al spins 2 and Carlos spins 3, Betty can win with either 6 or 8. In either case, she is simply credited with a win.)

○ Part I: Comparing the Spinners

To begin, investigate the outcomes of each of the possible two-person competitions in this game.

1. Suppose Betty plays directly against Al (without Carlos participating). What percentage of the time will Betty win? What percentage of the time will Al win? Explain your answers.

2. Suppose Betty plays directly against Carlos (without Al participating). What percentage of the time will Betty win? What percentage of the time will Carlos win? Explain your answers.

3. Suppose Al plays directly against Carlos (without Betty participating). What percentage of the time will Al win? What percentage of the time will Carlos win? Explain your answers.

continued

If you worked carefully, you found a rather strange combination of results. It turns out that Al beats Betty most of the time, so we can say that Al's spinner is better than Betty's. Similarly, Betty's spinner is better than Carlos's, because Betty beats Carlos most of the time. And Carlos's spinner is better than Al's, because Carlos beats Al most of the time.

If you didn't get these outcomes, reexamine your work. Keep in mind that for Questions 1 to 3, you need to give the percentage of the time each player wins, not simply say who wins most of the time.

○ Transitive Relationships

Why are these results strange? You might expect that because Al's spinner is better than Betty's and Betty's spinner is better than Carlos's, then Al's spinner must be better than Carlos's. But in fact, it isn't.

Many forms of comparison do work the way you'd expect. For example, suppose person A is taller than person B, and person B is taller than person C. You can then be sure person A is taller than person C. We express this by saying that "being taller than" is a *transitive relationship*.

The relationship "being greater than" (for real numbers) is also a transitive relationship. In other words, if $x > y$ and $y > z$, then you can be sure that $x > z$.

In Questions 1 to 3, you saw that for these spinners, "being better than" is not transitive. We can describe the spinners in this problem as "a nontransitive set of spinners."

continued ▶

○ *Part II: Investigating Transitivity*

The main task of this POW is to investigate some questions about nontransitive sets of spinners. Try to prove any conclusions you reach. Here are some questions to consider. If you get stuck on one question, try another. Also feel free to investigate other questions about transitive spinners.

• Is there *another* nontransitive set of three spinners, each divided into three equal-size parts, using each of the numbers 1 to 9 exactly once?

• Is there a nontransitive set of three spinners, each divided into two equal-size parts, using each of the numbers 1 to 6 exactly once?

• Is there a nontransitive set of three spinners, each divided into four equal-size parts, using each of the numbers 1 to 12 exactly once? If so, is there more than one such set of spinners?

• How do the answers to these questions change if you allow the same number to be used on more than one spinner—that is, if you make it possible for ties to occur?

○ *Write-up*

Your write-up should consist of your answers to Questions 1 to 3, with explanations, and any results you found for Part II (with proofs, if possible).

Mystery Tables

The table shown here actually represents In-Out tables for six different "mystery" functions, called f, g, h, F, G, and H. All six In-Out tables use the same set of x-values. To get the In-Out table for a given function, combine the first column of the table with the column for that function. For instance, the table shows that $G(-2) = -8$.

For each function, do these two things:

• Decide which family the function belongs to.

• Find an algebraic expression for the specific function.

As you work on each function, keep notes on what you tried and what led you to the next step. This will serve as your "detective's notebook" as you track down each function.

x	$f(x)$	$g(x)$	$h(x)$	$F(x)$	$G(x)$	$H(x)$
-5	26	35	27	0.015625	-230	96
-4	17	24	22	0.03125	-112	48
-3	10	15	17	0.0625	-42	24
-2	5	8	12	0.125	-8	12
-1	2	3	7	0.25	2	6
0	1	0	2	0.5	0	3
1	2	-1	-3	1	-2	1.5
2	5	0	-8	2	8	0.75
3	10	3	-13	4	42	0.375
4	17	8	-18	8	112	0.1875
5	26	15	-23	16	230	0.09375

Brake! Revisited

Let's now return to the table you saw in *Brake!*, which is reproduced here. It shows the distance a particular car travels in terms of the car's speed at the time the driver applies the brakes.

Speed (in miles per hour)	Stopping distance (to the nearest tenth of a foot)
20	22.2
25	34.7
30	50.0
35	68.0
40	88.8
45	112.4
50	138.8

You can now apply what you've recently learned about tables.

1. This table can be approximated very closely by a function from one of the basic families. Which family do you think this function belongs to? Justify your answer.

2. Find the family member. That is, find an equation, $y = f(x)$, where x stands for the speed of the car and y is the approximate stopping distance. Once you have an equation, give a careful description of how you found it.

Bigger Means Smaller

Questions 1 and 2 involve fairly straightforward situations. As you work on them, think about what principles they illustrate.

1. A farmer decides to devote an acre of his land to corn. (One acre is equal to 43,560 square feet.) Assume he decides to use a rectangular plot of land for the corn.

   a. If the width of the cornfield is 200 feet, what is the length (in feet)?

   b. How would the length have to change if the width were doubled? Tripled? Halved?

   c. Develop an equation for the length of a one-acre rectangular field in terms of its width.

2. Suppose you go on a 300-mile car trip.

   a. If your average speed is 50 miles per hour, how long will the trip take?

   b. How would the time of the trip be affected if your average speed were cut in half? What if you took a high-speed train that went twice as fast? Three times as fast?

   c. Develop an equation for the time of a 300-mile trip in terms of the average speed.

3. What is the mathematical connection between Questions 1 and 2?

4. Make up a situation and questions of your own that illustrate the ideas used in Questions 1 and 2.

Going to the Limit

In the next part of this unit, you will examine the fact that for some functions, certain numbers can't be used as inputs. You'll look at what this means in terms of the graphs of these functions when the inputs are near those "forbidden" values.

Mayleiya Henry reflects on what asymptotes mean in the context of specific problems.

Don't Divide That!

In *Bigger Means Smaller,* the equations you found probably involved fractions with a variable in the denominator. Such algebraic fractions raise some important questions.

- How do you evaluate an algebraic fraction when the denominator comes out to zero?
- What happens in the graphs of functions that have variables in the denominator?

In this activity, you will explore these two issues.

Part I: Why Can't You Divide by Zero?

Division by zero is said to be *undefined.* This seems strange to some people, who may reason like this: "You can divide by any other number, and zero is simply a number, so why can't you divide by zero?"

One answer to this question comes from the relationship between multiplication and division. For instance, we know that $12 \div 3 = 4$ because $4 \cdot 3 = 12$.

Discuss with your group how allowing division by zero would contradict this relationship between multiplication and division.

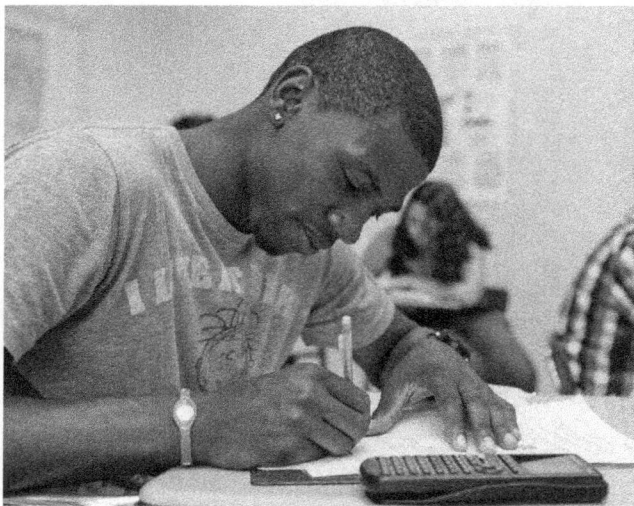

continued ▶

Part II: Graphs Involving Division by Zero

In Questions 1 and 2, a specific value of x makes the right side of the equation undefined because the denominator is zero at that x-value. This means there is no point on the graph for that x-value.

For each equation, do these three things:

- Find the value of x that will make the denominator equal to zero.
- Sketch a graph of the function by hand, considering all four quadrants. You will probably need to plot quite a few points to get a good sketch. The graphs might do funny things right around the x-values where the functions are undefined, so be sure to check x-values that are very close to those x-values.
- Graph the function on your calculator, and compare the calculator graph with your sketch.

1. $y = \frac{1}{x}$

2. $y = \frac{10}{3 - x}$

Difficult Denominators

You have seen that algebraic fractions create special complications for graphing.

For each function below, first make a table of values. Be sure to include *x*-values very near the vertical **asymptotes.** Then make a graph based on your table.

Plot each function on a separate set of axes. Give special attention to what happens to the graph near any vertical asymptotes. As in *Don't Divide That!,* you will probably need to plot quite a few points to get a clear idea of what the graph looks like.

1. $y = \dfrac{1}{x + 5}$

2. $y = \dfrac{1}{(x - 2)^2}$

3. $y = \dfrac{25x}{x^2 - 25}$ (*Note:* There are two values of *x* where this function is undefined.)

Return of the Shadow

A group of young boys and girls are standing near a lamppost. Ethan points to his little sister, Emily, and says, "My mom is twice as tall as Emily. If my mom were standing where Emily is, my mom's shadow would be twice as long as Emily's shadow."

Ethan apparently thinks that shadow length is directly proportional to height. Use diagrams and ideas from geometry to investigate whether this is true.

More specifically, consider the case of a lamppost that is 20 feet tall and a person who is standing 12 feet away from the lamppost. The length l of that person's shadow will be a function of his or her height h.

Find an expression for l as a function of h. In particular, find out whether l is directly proportional to h.

An Average Drive

Amparo takes a trip on a warm, sunny day to her favorite spot overlooking the ocean. The spot is 100 miles from her home. Amparo knows that to be home on time, she needs to average 50 miles per hour for her overall trip.

1. As Amparo reaches her destination, she realizes that she has made good time while still staying under the speed limit. She figures out that she averaged 60 miles per hour on her trip to the ocean. What should Amparo's average speed be on the return trip so that her average speed for the round trip will be 50 miles per hour?

   (*Warning:* The answer is not 40 miles per hour!)

2. On another trip to the same spot, Amparo drives at a more leisurely pace. When she reaches her destination, she realizes that she is late! She averaged only 25 miles per hour getting to the ocean. What should Amparo's average speed be on the return trip so that her average speed for the round trip will be 50 miles per hour?

   (*Warning:* The answer is not 75 miles per hour!)

3. Amparo makes this round trip frequently, and her goal is to average 50 miles per hour for the round trip. However, her average speed for the trip to the ocean varies.

   Let x be her average speed (in miles per hour) on the way to the ocean. Find an expression in terms of x that tells Amparo what her average speed should be on the return trip in order for her average speed for the whole trip to be 50 miles per hour.

Approaching Infinity

In *An Average Drive,* you examined the function defined by the equation $y = \frac{100x}{4x - 100}$ (or something equivalent) and saw that the graph of this equation has a horizontal asymptote at $y = 25$. That is, as x gets very big, the y-value of the function gets closer and closer to 25.

The behavior of a function as x gets very big is sometimes called its *end behavior*. Keep in mind that this term refers to both the positive and negative ends of the x-axis.

When x is increasing arbitrarily (that is, going "far" in the positive direction), we say, "x is approaching positive infinity." We write $x \to +\infty$ (or simply $x \to \infty$) as shorthand for this.

If the value of the function $f(x)$ is getting very close to some number as x gets very large, we represent that number by the notation $\lim_{x \to \infty} f(x)$. This expression is read as "the limit of $f(x)$ as x approaches infinity." For example, for very large values of x, the expression $\left(1 + \frac{1}{x}\right)^x$ gets close to the special number e, so $\lim_{x \to \infty} \left(1 + \frac{1}{x}\right)^x = e$.

As the value of x moves toward the negative end of the x-axis, we say, "x is approaching negative infinity" and write $x \to -\infty$. If the value of $f(x)$ is getting very close to some number as x gets "very large" in the negative direction, we represent that number by the notation $\lim_{x \to -\infty} f(x)$. If the same thing is happening whether $x \to \infty$ or $x \to -\infty$, we can write $|x| \to \infty$. (The phrase "getting very close to" is being used here in an intuitive way. The formal mathematical definition is quite technical.)

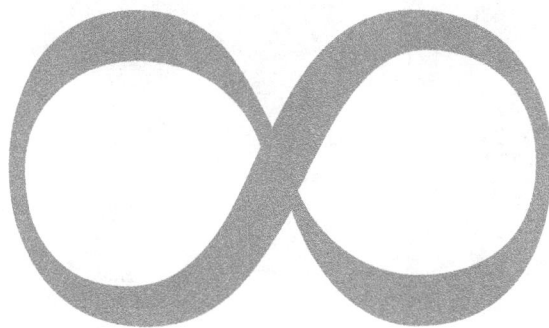

The End of the Function

You have seen that horizontal asymptotes are a particular type of end behavior. But what else can happen with end behavior? Keep in mind that a graph can have two "ends": one end where x is a large positive number and the other end where x is a "large" negative number (that is, a negative number with a large absolute value).

1. Look at examples of **polynomial functions** of different degrees. In each case, try to figure out what happens to the y-value as x increases in absolute value. (The result may depend on whether x is positive or negative.) Take notes on your results and any general conjectures you make.

   A calculator can help you to some extent, but the graph a calculator shows can give you only part of the picture. So you should come up with algebraic or numeric explanations for your conclusions about the end behavior of your examples.

2. Now look at functions from other families: exponential functions, functions from the sine family, **rational functions,** and any other families you want to consider. Again, take notes on your results and any general conjectures you make.

Creating the Ending You Want

You have looked at specific functions and examined their end behavior. Now you will turn the process around.

1. Parts a to d describe functions with four possible end behaviors. For each behavior, do these three things:

   - Draw a graph of an equation with that behavior.
   - Create an algebraic equation for a specific function that behaves in that way.
   - Explain how you know that your equation leads to the given end behavior.

   a. A function for which $y \to \infty$ (that is, y becomes very large and positive) as $x \to \infty$ and for which $y \to -\infty$ (that is, becomes "large negative") as $x \to -\infty$

   b. A function for which $y \to \infty$ both as $x \to \infty$ and as $x \to -\infty$

   c. A function for which $y \to \infty$ as $x \to \infty$ and that has the x-axis as a horizontal asymptote as $x \to -\infty$

   d. A function with the x-axis as a horizontal asymptote at both ends

2. List some other possible end behaviors. If you can, find equations for functions with those behaviors.

Who's Who?

You have become pretty good at finding which family certain functions belong to. Now you will learn ways of identifying a particular family member to suit a given situation and set of data.

Jessica Harris and Victor Hernandez work on the graphs and equations they used to fit specific criteria before presenting them to the class.

Families Have Many Different Members

One step in finding a function to fit a situation is deciding which family the function should be in. Even after you've identified the family, the problem is not completely solved. You still have to figure out which *member* of the family it is. That usually means determining the special numbers, often called **parameters,** that distinguish the desired function from its "relatives." For instance, the parameters in the general linear function $y = ax + b$ are a and b, whereas x and y are the variables.

The situations in Questions 1 and 2 will give you experience with finding the values of some parameters.

1. For a science experiment in biology, Binh is observing the growth of a bacteria colony. At 2 p.m., he estimates there are 1000 bacteria. When he returns to check at 5 p.m., there are about 2200.

   a. If the bacteria continue to reproduce at this rate, how many will there be by midnight?

   b. Find a function that will tell you how many bacteria will be present at any given time.

2. A helicopter is bringing supplies to an isolated town where housing has been destroyed by an earthquake. As the helicopter hovers 200 feet above the ground, a crew member throws a pack of blankets downward with a certain force. Three seconds later, the blankets reach the ground. How high off the ground was the pack of blankets after 2 seconds?

   Reminder: If an object has initial height h_0 and initial upward velocity v_0 at time $t = 0$, its height off the ground (in feet) after t seconds is given by the expression $h_0 + v_0t - 16t^2$. An initial downward velocity is represented by a negative value of v_0.

Fitting Mia's Birdhouses Again

Do you remember Mia and her birdhouses? They initially appeared in the Year 1 unit *The Pit and the Pendulum*. In that unit, you were told how many birdhouses Mia and her friends had painted after 1 hour and after 3 hours. Your task was to predict how many they would paint in 8 hours.

Mia and her friends reappeared in the Year 3 unit *Meadows or Malls?* in a problem that provided additional information. You had to explain why a linear function wouldn't fit the data.

Perhaps you don't remember the details, so here they are:

> Mia and her friends spent the semester building birdhouses, and now they are painting them. In 1 hour, they painted two birdhouses. After 3 hours, they had painted six birdhouses. After 5 hours, they had painted nine bird houses.

1. Plot this data set. Use number of hours for the *x*-axis and number of birdhouses painted for the *y*-axis.

2. Find a quadratic function that fits this data set. That is, find specific numbers for *a*, *b*, and *c* so that all three data points fit the equation $y = ax^2 + bx + c$.

3. Does it make sense to use a quadratic function to fit this data set? Why or why not? Are there any other function families you think might better fit this situation and data set?

Mystery Tables II

As in *Mystery Tables,* the tables here each combine In-Out tables for several functions. Your task again is to find an algebraic equation for each function.

x	f(x)	g(x)	h(x)
−5	−7	20	−7
−4	−5	18	−13
−3	−3	16	−17
−2	−1	14	−19
−1	1	12	−19
0	3	10	−17
1	5	8	−13
2	7	6	−7
3	9	4	1
4	11	2	11
5	13	0	23

x	F(x)	G(x)	H(x)
0	0	0	2
15	0.26	0.5	2.26
30	0.50	0.87	2.50
45	0.71	1	2.71
60	0.87	0.87	2.87
75	0.97	0.5	2.97
90	1	0	3
105	0.97	−0.5	2.97
120	0.87	−0.87	2.87
135	0.71	−1	2.71
150	0.50	−0.87	2.50
165	0.26	−0.5	2.26
180	0	0	2

The table on the left represents functions *f, g,* and *h* and uses one set of input values. The table on the right, for functions *F, G,* and *H,* uses a different set of input values.

For each of the six functions, do these three things:

- Graph each function on its own pair of axes, with appropriate scales.
- Decide which family the function belongs to, and explain your reasoning.
- Find a specific algebraic equation for the function.

What Will It Be Worth?

Imagine that you bought a house in the year 2000 for $100,000 and took out a 30-year mortgage as part of the purchase process. This means you borrowed money to buy the house and will pay back the money over a 30-year period.

As part of your financial planning, you would like to make an educated guess about how much your house will be worth when the loan is paid off, in 2030. You know the house was worth $50,000 in 1980.

1. First, assume housing prices are increasing linearly over time. Based on this model, what is your prediction?

2. Next, assume housing prices are increasing exponentially over time. Based on this model, what is your prediction?

3. Now pick a third model. That is, come up with some other possible way in which prices might be changing. Then give your prediction using this model, and explain your reasoning.

4. Which of the three models do you think makes the most sense? Why?

The Decision About Dunkalot

The general manager of the Slamajamas basketball team has a difficult decision to make. A key player, Marcus Dunkalot, suffered a sprained knee on March 20, about one month before the playoffs, and was put on the disabled list. The playoffs begin on April 18.

It is now April 6, and the general manager has just received the physical therapist's report, shown here.

PROFESSIONAL PHYSICAL THERAPY

Patient's name: Marcus Dunkalot

Sex: male   Height: 6′8″

Age: 24   Weight: 225 lb

Diagnosis: sprained knee

Prescribed treatment: strengthen and stretch

3/20	Mr. Dunkalot was administered a Cybex strength test upon arrival. Quadriceps of the injured leg measured 55 foot-pounds in extension. Normal measurement for a player to return to play without reinjury is 250 foot-pounds.
3/25	Daily regimen is contributing to patient's progress. Cybex test measures 90 foot-pounds.
4/1	Some swelling earlier in the week. General reports of less pain. Cybex test measures 140 foot-pounds.
4/6	Less swelling. Range of motion shows marked increase. Cybex test measures 185 foot-pounds.

The manager needs to decide immediately whether to keep Marcus on the disabled list. If he keeps Marcus on the list, it will be for the remainder of the regular season. That means Marcus will be disqualified for the playoffs.

continued ▶

Here are the advantages and disadvantages of each choice:

- If he takes Marcus off the list now, he can hope Marcus will be well in time for the playoffs. But if Marcus isn't ready in time, the Slamajamas will have one less player available for the rest of the season, including the playoffs.

- If he keeps Marcus on the list, then he can sign another player (of lesser ability) to take his place. But he gives up all hope of having Marcus for the playoffs.

Two Suggested Solutions

Two students working on this problem give the coach different advice. Both students designated March 20 as Day 1, so April 18 became Day 30. (March has 31 days.)

Student A says the function f given by the equation $f(x) = 40 + 8x$ approximates the data set well. Student A predicts that on April 18, Marcus will have 280 foot-pounds of strength and will be strong enough to play.

Student B says the function g given by the equation $g(x) = 55 + 6x$ approximates the data set well. Student B predicts that on April 18, Marcus will have only 235 foot-pounds of strength and will not be strong enough to play.

Your Questions

1. Which student's function seems to you to fit the data set better? Why?

2. Do you have a function that you think fits the data set better than either of these? If so, what is it?

3. Develop a mathematical procedure by which you might judge when one function fits a data set better than another.

A Tight Fit

In this and previous units, you have often tried to find a function whose graph came very close to a given set of data points. Now you will examine the question of what "close" means.

Daniel Smith uses his calculator's regression feature to further explore the meaning of "best fit."

Let's Regress

At various times in your study of mathematics, you have done some curve-fitting. If you know the family of functions a data set belongs to, your calculator can use a technique called *regression* to find the function of best fit for that family.

When you can solve problems that way, life is easy. In this activity, you apply regression in two situations.

Part I: The Hobby Shop

Steve works in a hobby shop. He was looking at the relationship between the length of certain wooden models and the amount of paint they require. Here are some of the estimates he came up with.

Length of model (in feet)	Amount of paint (in ounces)
1	2
2	6
3	14
5	40

Assume this pattern continues. Your task is to use your calculator's regression feature to estimate how much paint would be needed for a model that is 10 feet long.

1. Choose a regression model. That is, choose a function family to model the situation. Explain your choice.

2. Apply your regression model to the data in the table, and give the resulting function.

continued ▶

Part II: Growing Up

This graph shows the average height for boys ages 0 to 6 in the United States around the middle of the twentieth century. You may recall the graph from an activity in the Year 3 unit *Small World, Isn't It?*

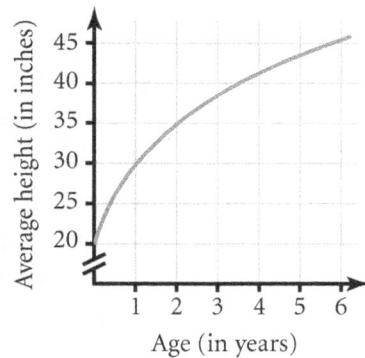

3. Suppose you want to use regression with the information in this graph to predict the average height for a 9-year-old boy in the United States at the time.

   a. Choose three different models for regression, and explain your choices.

   b. Give the function and the prediction you get for each model.

   c. Discuss which model you think is best, and explain your reasoning.

Midnight Express

1. The Midnight Express postal train moves along at a steady 70 miles per hour, passing through Nowheresville exactly at midnight, without even slowing down.

   a. Make a graph showing the train's distance from Nowheresville as a function of time. Use midnight as $t = 0$, and include both positive and negative values of t in your graph. For example, $t = -1$ will represent an hour before midnight, or 11 p.m.

   b. Write a rule for how to find the train's distance from Nowheresville as a function of t.

2. This table gives first-class mail rates in 2010 for large envelopes.

Weight	Cost
1 ounce or less	88¢
More than 1 ounce but no more than 2 ounces	105¢
More than 2 ounces but no more than 3 ounces	122¢
More than 3 ounces but no more than 4 ounces	139¢

   The rates continue like this up to 13 ounces, with each additional ounce costing another 17¢.

   a. Make a graph showing cost as a function of weight, up to a weight of 7 ounces.

   b. Write a rule for how to find the cost of mailing a large envelope that weighs n ounces.

It's Off to College We Go

Your task in this POW is to investigate and compile information on the cost of going to college for four years. The planning and mathematics you do will serve you well whether you actually attend college or not.

Keep in mind that college expenses include more than simply tuition, fees, and the cost of books. They also include ordinary living expenses such as food and housing.

Be very detailed and accurate in your plan. For instance, if you plan to live in a dormitory, find the cost of dormitory living at the college of your choice. If you want to get your own apartment, then research rentals in that area to get an idea of the cost.

You might plan to earn a partial scholarship or to get a part-time job during the school year. You will need to research how much money that will bring in. If you plan to work summers, determine how much you can earn from summer employment.

It will probably be helpful to interview people who are now in college—there is nothing like experience. What bills will you have to pay? How much should you plan to spend on books? On recreation and entertainment?

Also keep in mind that prices do not stay fixed from year to year. You've studied the way inflation works in a variety of activities in previous units. To be specific, plan on an annual 2.5% inflation rate. That is, once you find the cost for the first year, plan for prices to rise by 2.5% in each additional year.

○ Write-up

Write a report of your work and conclusions. Include a list of expenses for each year, as well as an explanation of how you arrived at those figures.

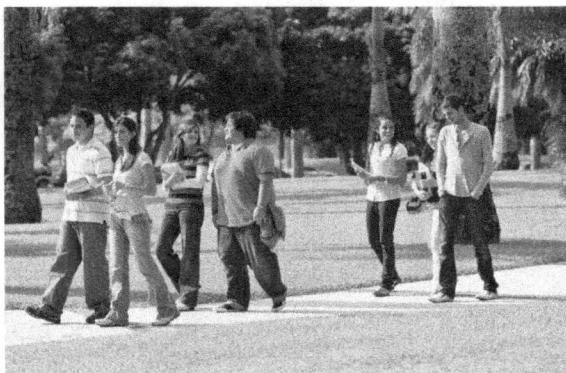

In the Lead

1. Our favorite track star, Speedy, has a friend named Sporty who is also quite athletic. Although running isn't Sporty's best event, the two friends like to race each other.

   You may recall this information about Speedy from the Year 3 unit *Small World, Isn't It?*

   > A sports analyst studied the video of one of Speedy's races. The analyst came up with the function $m(t) = 0.1t^2 + 3t$ to describe the distance, in meters, Speedy had run t seconds after beginning the race.

   The same analyst came up with the expression $0.095t^2 + 2.92t$ to describe the distance Sporty has run t seconds after beginning the race.

   a. After 10 seconds, how far ahead of Sporty is Speedy?

   b. Write a function that describes how far ahead Speedy is after t seconds.

2. On January 1, 1970, the population of Smallville was 30,000. Over the next decade, the population grew by 2% each year.

   a. Write an expression for the population of Smallville t years after January 1, 1970.

   During the period from 1970 to 1980, electricity usage rose as well. In fact, the usage per person went up. The average electricity usage per person as of January 1, 1970, was 5 kilowatts per day. On January 1, 1980, the average usage per person was 7 kilowatts per day.

   b. Assume the per-person electricity usage grew linearly during this period. Based on this assumption, find an expression for the average daily usage per person at time t.

   c. Use your answers to find an expression for the *total* daily electricity usage in Smallville at time t.

Back to Arithmetic

You may have thought you were finished studying arithmetic a long time ago. Now you will return to that study, but not to the arithmetic of numbers. You will learn how to use arithmetic to create new functions from existing ones.

Leslie Gonzalez and Marissa McMahan listen to a presentation on "Back to the Corral."

The Arithmetic of Functions

We can do arithmetic with functions almost as easily as we do arithmetic with numbers or algebraic expressions. For example, if f and g are any two functions, we can define a new function, called $f + g$, by the equation

$$(f + g)(x) = f(x) + g(x)$$

Suppose the function f is defined by the equation $f(x) = x^2 + 3x + 4$ and the function g is defined by the equation $g(x) = 2x^2 - x + 1$. Then we have

$$(f + g)(x) = (x^2 + 3x + 4) + (2x^2 - x + 1)$$

In other words, $f + g$ is the function defined by the equation $(f + g)(x) = 3x^2 + 2x + 5$.

We can define the functions $f - g$, $f \cdot g$, and $f \div g$ similarly. We can even create more complex combinations out of f and g, such as defining $3f - g$ by the equation

$$(3f - g)(x) = 3(x^2 + 3x + 4) - (2x^2 - x + 1)$$

Functions Defined by Tables

Arithmetic operations can also be defined when the original functions are given in terms of a table. In this case, the new functions are defined only for numbers that are in the **domain** for both functions.

For example, here are In-Out tables for two functions h and k.

x	$h(x)$		x	$k(x)$
1	-4		2	7
2	5		3	-5
3	-8		4	6
5	2		5	1
8	2		7	-3

continued

Your Task

Use the information in the In-Out tables for h and k to make In-Out tables for each of these functions. Make your tables as complete as possible.

1. $h + k$

2. $h \cdot k$

3. $2h - k$

4. $h^2 + k^2$

5. $(h + 1)(k + 2)$

The Arithmetic of Graphs

You've explored some problems in which the function you were looking for could be expressed by doing some arithmetic operation on simpler functions.

Now you will look at how this "arithmetic of functions" can turn into an "arithmetic of graphs." You may want to use colored pencils so you can easily identify the graphs of different functions.

1. Define two functions, h and k, by the equations $h(x) = 2x - 1$ and $k(x) = x + 2$.

   a. On a single set of axes, draw the graphs of h and k between $x = -3$ and $x = 3$.

   b. Draw the graph of the function $h + k$ on the same set of axes.

   c. Use the cases $x = -3$, $x = -1$, and $x = 2$ to describe how points on the graph of $h + k$ can be found from the graphs of h and k. Try to express the process in terms of the lengths of certain line segments.

2. The graphs of two functions, f and g, are shown on the axes in this diagram. Make a copy of this pair of graphs. Then, on the same set of axes, draw a graph for each of these combination functions.

   a. $f + g$

   b. $f - g$

   c. $2g$

3. You can't use the graphs in Question 2 to get the graph of the product of the two functions, $f \cdot g$. Why not?

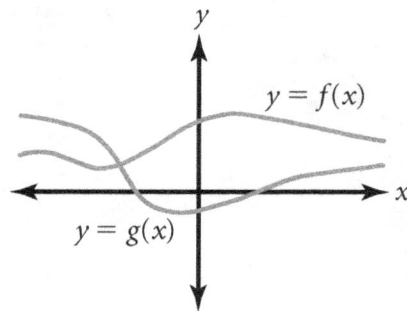

Back to the Corral

Do you remember Rancher Gonzales and her corrals from the Year 2 unit *Do Bees Build It Best?* Rancher Gonzales knew how much fencing she had, and her goal was to build the corral with the largest possible area using that amount of fencing.

Now she has a slightly different problem. She knows how much area she wants, and her goal is to minimize the amount of fencing. Specifically, she has two new horses and wants to build a corral for them with an area of 300 square feet. She will need to decide on the exact shape of her corral.

1. Rancher Gonzales's initial plan is to build a rectangular corral. She needs to determine the dimensions that will give an area of 300 square feet while using the least possible amount of fencing.

   a. Before working it out, make a guess about what the dimensions will be.

   b. Demonstrate why your answer is correct.

2. Nephew Juan is back on the scene. Two years ago, he thought rectangles were boring, and he still does.

This time he wants his aunt to build a corral in the shape of a rectangle with a semicircle on one end, as illustrated in the diagram. The fence will go only around the outside. The dashed line is shown so you can identify the semicircle. No fencing is needed where the dashed line is.

Determine what dimensions will give the corral an area of 300 square feet and keep the amount of fencing at a minimum. Justify your answer.

Name That Family!

In each of the situations in Questions 1 to 3, one variable is described as a function of another. For each function, do these three things:

- Sketch a graph.
- State which family the function belongs to.
- Explain your reasoning, including any assumptions you make about the situation.

If you don't think the function fits exactly into any standard family you know, indicate which family you think comes closest.

1. A group of friends is weeding a garden. In this situation, the time needed to do the job is a function of the number of people weeding.

2. A hot cup of coffee is placed on a table and forgotten. In this situation, the temperature of the coffee is a function of time. (*Note:* The rate at which the coffee's temperature decreases is proportional to the difference between the temperature of the coffee and the temperature of the room.)

3. An ice cream store has a collection of cones, some larger and some smaller but all the same shape. The ratio of the diameter at the top of a cone to the height of the cone is the same in all cases.

   In this situation, the amount of ice cream a cone holds is a function of the cone's height. Assume cones are always filled level to the top, but no higher.

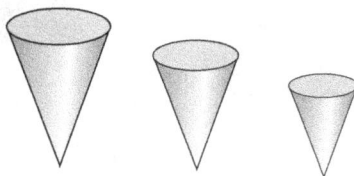

Small World, Isn't It? Revisited

The rate of world population growth has varied over the centuries. This makes it difficult to get a good fit to all the available data using an exponential function. But if you look at only some of the data for a recent period, you might be able to get better results.

Consider these world population estimates, which are part of the data set you used in the Year 3 unit *Small World, Isn't It?*

Year	Estimated population
1960	3,040,000,000
1980	4,450,000,000
2000	6,090,000,000

1. Using this table or a graph made from the data, what would you estimate the population to be in the year 2060?

2. Find a general expression for predicting the world population in some arbitrary future year, based on this data set.

3. What population does your expression predict for the population in the year 2060? How does this compare with your estimate?

Composing Functions

You know that people compose music and poetry. Did you know they also compose functions? You will soon learn about an operation called *composition* that provides an interesting way of creating new functions from existing ones.

What you discover about the **composition of functions** in the next several days will serve you well if you ever study calculus.

Shafari Jackson graphed several functions in order to become familiar with the composition of functions.

Rumble, Grumble

When David is hungry, he gets grouchy. When he isn't hungry, he stays in a good mood. In other words, David's mood is a function of his hunger level.

David's friend Karla usually has positive feelings toward him. But when he is grouchy, Karla gets irritated with him. In other words, Karla's attitude toward David is a function of his mood.

Assume David eats three meals a day, at regular times, so that his hunger level is a function of the time of day.

Sketch a graph of Karla's feelings toward David as a function of time over the course of two days.

The Composition of Functions

You have discovered that you can combine functions using arithmetic operations. You can also create new functions by first applying one function and then applying another.

Given two functions, f and g, we can combine them to create a new function called the *composition* of f and g. This new function is written as $g \circ f$, and its output for the input x is $g(f(x))$. Thus, $g \circ f$ is defined by the equation $(g \circ f)(x) = g(f(x))$. A new function created in this way is called a *composite function*.

You can visualize the composition of functions using a function machine like this one. It shows x as the input for the function f. Then it shows the output $f(x)$ as the input for the function g.

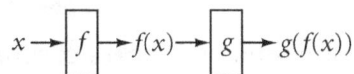

$$x \longrightarrow \boxed{f} \longrightarrow f(x) \longrightarrow \boxed{g} \longrightarrow g(f(x))$$

The situations in Questions 1 and 2 illustrate the operation of composition.

1. Maria is working to save for college. This graph shows her earnings over a period of five weeks as a function of the week.

   Each week, Maria gives $50 of her earnings to her mother and saves half of the rest. The amount she saves is a function of her earnings.

   Draw a graph showing the amount Maria saves each week.

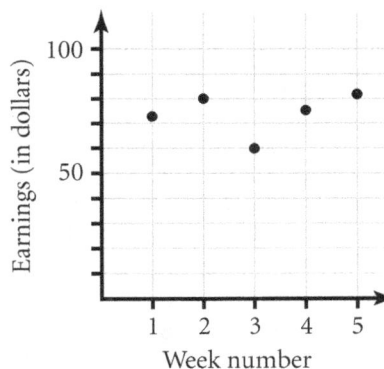

2. Mario is also saving for college. He wants to earn enough money during his senior year of high school to put aside $500 for college expenses and have another $500 to travel with friends across the country after graduation.

continued

Mario decides to sell magazine subscriptions. Of course, he earns more if he sells more subscriptions, but it's better than that. He earns more *per subscription* if his sales go up. The chart shows how much he earns for each subscription sold in a certain month.

Subscription number	Earnings for each subscription
1st through 10th	$5 each
11th through 20th	$7.50 each
21st through 30th	$10 each
31st and beyond	$15 each

For example, if Mario sells 7 subscriptions in a month, he gets $5 for each of them, for a total of $35 for that month. But if he sells 14 subscriptions in a given month, he gets $5 for each of the first 10, and then $7.50 for each of the next 4, for a total of $80. And if he sells 23 subscriptions in one month, he gets $5 for each of the first 10, $7.50 for each of the next 10, and $10 for each of the last 3, for a total of $155.

a. Make either a graph or a table of Mario's earnings in a given month as a function of the number of subscriptions he sells that month. Include earnings through at least 25 subscriptions.

b. The school year is now over. The next table shows the number of subscriptions Mario actually sold each month as a function of the month.

Make a table showing Mario's earnings each month.

c. Determine whether Mario will be able to take his cross-country trip.

Month	Number sold
September	7
October	15
November	9
December	32
January	8
February	13
March	23
April	18
May	36

The Cost of Pollution

Once there was a peaceful river that emptied into a clear lake. The lake was filled with catfish, and a wonderful restaurant on the shore of the lake featured delicious catfish dinners.

Then a chemical company decided to build a plant along the river, upstream from the lake. Instead of disposing of its toxic waste in a responsible manner, the company dumped the waste right into the river. You've probably guessed the rest. The catfish began getting sick and dying.

There are two functions involved in this situation. One function gives c, the number of catfish per acre, as a function of w, the number of gallons of toxic waste dumped per day. For this activity, assume that $c = 50 - 5w$. We will write this function as $c = G(w)$, so G is defined by the equation $G(w) = 50 - 5w$.

The second function expresses p, the price (in dollars) of a catfish dinner, as a function of c, the number of catfish per acre. Assume that $p = 24 - 0.4c$. We will write this function as $p = H(c)$, so H is defined by the equation $H(c) = 24 - 0.4c$.

1. Draw a function machine showing how the variables w, c, and p are related using the functions G and H.

2. Write an equation expressing p in terms of w. Represent this equation symbolically using composition notation.

continued ▶

3. What was the cost of a catfish dinner before the chemical company moved in?

4. Suppose you go to the restaurant and find that a catfish dinner costs $12. How much waste is the chemical company dumping per day?

5. How much waste will the company have to dump per day in order for all of the catfish to be gone?

6. Make up your own situation using the composition of functions. You might want to use an ecology theme.

Adapted from page 45 of *Earth Algebra* by Christopher Schaufele and Nancy Zumoff (New York: Addison Wesley Longman, 1993). Reprinted by permission of Addison Wesley Longman (Pearson Education).

Order Among the Functions

You learned long ago that for some numeric operations, you could interchange the numbers without changing the result. For example, the expressions $5 + 7$ and $7 + 5$ both give the sum 12. This fact is expressed by saying that addition is *commutative*.

You also know that for other operations, the order in which numbers appear is important. For instance, $12 - 8$ is not the same as $8 - 12$. In other words, subtraction is *noncommutative*.

You will now consider several questions related to the commutativity or noncommutativity of the operation of composition of functions.

1. Define functions f and g by the equations $f(x) = x + 3$ and $g(x) = 7x$.

   a. Create separate function machines for both $f \circ g$ and $g \circ f$.

   b. Find an equation for the composite function $f \circ g$. That is, find an expression for $f(g(x))$.

   c. Find an equation for the composite function $g \circ f$.

   d. Is $f \circ g$ equal to $g \circ f$? In other words, do these two composite functions give the same output for every possible input?

2. Now explore whether there may be cases different from your finding in Question 1d.

   a. Find a function h for which the composite functions $h \circ f$ and $f \circ h$ are equal. Find several choices for h if you can.

   b. Find a function k for which the composite functions $k \circ g$ and $g \circ k$ are equal. Find several choices for k if you can.

Cozying Up to Composition

In Part I of this activity, you will find numeric values or algebraic equations for various composite functions. In Part II, you will look at a familiar function and find two functions whose composition is that familiar function.

Part I: Putting Functions Together

The functions f, g, h, and k are defined by these equations:

- $f(x) = 3x^2 + 5$
- $g(x) = 2x - 7$
- $h(x) = 2^x$
- $k(x) = \sin x$

1. Find the numeric value of each expression.
   a. $(g \circ f)(4)$
   b. $(f \circ g)(-2)$
   c. $(h \circ f)(1)$
   d. $(g \circ k)(90°)$

2. Find an algebraic equation for each composite function. Simplify your equations if possible.
   a. $g \circ f$
   b. $f \circ g$
   c. $f \circ k$
   d. $h \circ g$
   e. $f \circ f$
   f. $g \circ g \circ g$

continued ▶

Part II: Revisiting a Mess!

You may have met pilot Lindsay in the Year 3 unit *Small World, Isn't It?* She flew over an oil slick that was in the shape of a circle. When Lindsay first flew over the circular oil slick, its radius was 70 meters. She estimated that the radius of the circle was increasing at a rate of 6 meters per hour.

3. Let t represent the number of hours since Lindsay first saw the oil slick, and let $A(t)$ be the area of the oil slick after t hours. Write an equation for $A(t)$ in terms of t.

4. Define two specific functions whose composition is the area function $A(t)$ from Question 3.

Taking Functions Apart

In Part I of *Cozying Up to Composition*, you put functions together to create more complex functions, using the operation of composition. Now you will do the opposite. You will start with a complex function and find the simple functions that make it up.

Often, you can see how to do this by examining how you would compute a numeric output for the function. For example, consider the function defined by the equation $h(x) = (2x + 3)^3$. To find $h(5)$, you can first compute $2 \cdot 5 + 3$, getting 13, and then find 13^3. To express h as the composition of two functions, you can define g by the equation $g(x) = 2x + 3$ and f by the equation $f(x) = x^3$ and then write h as the composite function $f \circ g$.

Each expression in Questions 1 to 6 defines a function. Write each of these functions as the composition of two or more other functions. Give clear definitions for the functions you are using. There is more than one correct answer for some of these problems.

1. $3x^2$

2. $\dfrac{1}{\sin x}$

3. $(3x^2 - 6)^5$

4. $\sqrt{x^2 + 2x - 4}$

5. $\ln (x^2 + 1)$

6. $3^{(x+2)} + 7$

Fish, Ladders, and Bacteria

As you work on these problems, think about what they have in common.

1. In *The Cost of Pollution*, you were told that c, the number of catfish per acre, could be expressed in terms of w, the amount of waste dumped per day, by the equation $c = 50 - 5w$.

   a. If there are 20 catfish per acre, how much waste is being dumped per day?

   b. Develop a general expression for the amount of waste being dumped per day in terms of the number of catfish per acre.

2. A 12-foot ladder leans against a house, making an angle θ with the ground. The height of the top of the ladder, h, is given by the equation $h = 12 \sin \theta$.

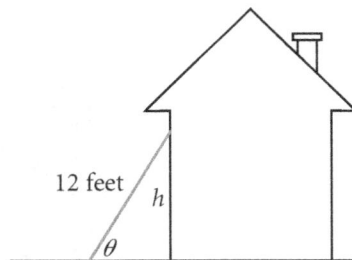

   a. If you want the top of the ladder to be 11 feet off the ground, what should you use for the angle between the ladder and the ground (to the nearest degree)?

   b. Develop a general expression for the angle between the ladder and the ground in terms of the height of the top of the ladder.

3. Suppose b, the number of bacteria in a colony, starts at 50 and doubles every hour. After t hours, the value of b can be expressed by the equation $b = 50 \cdot 2^t$.

   a. How much time has gone by at the moment the bacteria population reaches 1000? Give your answer to the nearest tenth of an hour.

   b. Develop a general expression for the amount of time gone by in terms of the size of the bacteria population.

Functions in Verse

The **inverse of a function** "undoes" the original function. For example, if f is the function defined by the equation $f(x) = x - 8$, then its inverse function, written as f^{-1}, can be defined by the equation $f^{-1}(z) = z + 8$. (It's often helpful to use a different independent variable for the inverse function than was used for the original function.) Similarly, if $k(t) = t^3$, then $k^{-1}(r) = \sqrt[3]{r}$.

In this activity, you will look at how the table and the graph of a function are related to the table and the graph of its inverse.

1. Start with the function $k(t) = t^3$.

   a. Complete the In-Out table for k shown here. Note that this table does not show all possible values of x.

x	$k(x)$
-1	
2	
4	
-3	
5	

   b. Make an In-Out table for k^{-1}. Use the numbers in the *Out* column from part a as the inputs for k^{-1}.

2. Now consider the function $f(x) = x - 8$.

   a. Make an In-Out table for f, using at least five different input values.

   b. Make an In-Out table for f^{-1}, using the numbers in the *Out* column from part a as the inputs.

continued

3. This table is part of an In-Out table for some function h. Use this information to make an In-Out table for h^{-1}.

x	$h(x)$
3	7
5	9
7	6
9	15
11	13

4. Based on Questions 1, 2, and 3, state a general principle about the relationship between the table of values for a function and the table of values for its inverse. Explain why your principle is true.

5. a. Sketch a graph for the function k (from Question 1) using the same scale for the x- and y-axes. Then sketch a graph of k^{-1} on the same set of axes.

   b. Sketch a graph for the function f (from Question 2) using the same scale for the x- and y-axes. Then sketch a graph of f^{-1} on the same set of axes.

   c. Plot the information you have from Question 3 for the function h, using the same scale for the x- and y-axes. On the same set of axes, plot the points you know for the function h^{-1}.

   d. This diagram shows the graph of some function g. Assume the x- and y-axes use the same scale. Copy this graph, and then sketch the graph of g^{-1} on the same set of axes.

   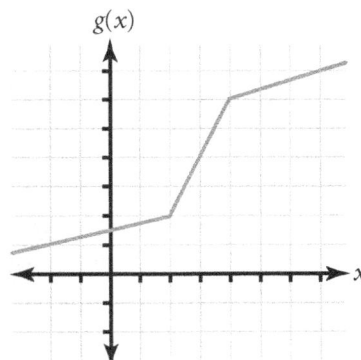

   e. Based on parts a to d, state a general principle about the relationship between the graph of a function and the graph of its inverse. Explain why your principle is true.

Linear Functions in Verse

You have been examining functions and their inverses. It's important to be aware, though, that not every function has an inverse. For example, the fact that $\sin 30° = \sin 150°$ means the **sine function** cannot have a true inverse. After all, $\sin^{-1} 0.5$ cannot equal both $30°$ and $150°$.

Even when a function has an inverse, it isn't always possible to find an equation for the inverse from the equation of the function itself. For instance, it's possible to show that the function defined by the equation $y = x^7 + 5x^3 + 6x$ has an inverse. However, there is no equation for this inverse function that expresses y in terms of x.

Fortunately, many basic functions do have inverses. In particular, not only does every nonconstant linear function have an inverse, but you can always find an equation describing the inverse.

You will now explore inverse functions for the family of linear functions.

1. Suppose f is the function defined by the equation $f(x) = 3x + 2$.
    a. Find an equation for the function f^{-1}.
    b. Make an In-Out table for the function f by using $-2, -1, 0, 1,$ and 2 as the *In* values and finding the corresponding *Out* values. Then check your answer to part a by applying that inverse function to each *Out* value in your table. If you don't get your *In* values as the results, you've made a mistake somewhere.

2. Generalize your algebraic work from part a. That is, suppose f is the linear function defined by the equation $f(x) = ax + b$ with $a \neq 0$. Find an equation for $f^{-1}(w)$ in terms of a, b, and w.

An Inventory of Inverses

Part I: Identity Elements

You are familiar with the concept of an identity element (or **identity**) for arithmetic operations. For example, the number 1 is the identity element for the operation of multiplication, because $1 \cdot x = x \cdot 1 = x$, for any number x.

There is also an identity element for the operation of composition of functions. It is the function I, called the **identity function,** that is defined by the equation $I(x) = x$. The statement that I is the identity element for composition means that $I \circ f = f$ and $f \circ I = f$ for every function f.

1. First you will find identity elements for several other operations.

   a. Find a number that acts as the identity element for addition of numbers. Explain your reasoning.

   b. Find a 3-by-3 matrix that acts as the identity element for addition of 3-by-3 matrices. Explain your reasoning.

   c. Find a 2-by-2 matrix that acts as the identity element for multiplication of 2-by-2 matrices. Explain your reasoning.

   d. *Challenge:* Define an operation ∗ on people in the world, like this:

      If A and B are any two people, then A ∗ B is the shorter of the two people. If A and B are identical in height, then A ∗ B = A.

      Describe a person in the world who would be an identity element for this operation. Explain your reasoning.

continued ▸

Part II: Inverses

Whenever an operation has an identity element, the next question to ask is whether individual elements have inverses. Intuitively, one element is the inverse of another if you get the identity element when you combine them using the given operation.

For example, the number $\frac{1}{5}$ is the inverse for 5 for the operation of multiplication because $\frac{1}{5} \cdot 5 = 1$ and $5 \cdot \frac{1}{5} = 1$ (and 1 is the identity element for multiplication).

Some functions have inverses under the operation of composition. Two functions f and g are inverses of each other if $f \circ g = I$ and $g \circ f = I$, where I is the identity function.

2. Consider the functions f and g defined by the equations $f(x) = 3x + 2$ and $g(x) = \frac{1}{3}(x - 2)$.

   a. Show the computation that demonstrates that $f \circ g = I$.

   b. Show the computation that demonstrates that $g \circ f = I$.

   c. Find an inverse (under composition) for the function k defined by the equation $k(x) = 4x^3 - 5$. Show how you know your answer is the inverse.

3. In Question 1, you found identity elements for several operations. Now consider the question of inverses for elements using those operations and their identity elements.

   a. Does every number have an inverse for addition? Explain your reasoning.

   b. Does every 3-by-3 matrix have an inverse under matrix addition? Explain your reasoning.

   c. Does every 2-by-2 matrix have an inverse under matrix multiplication? Explain your reasoning.

   d. *Challenge:* Consider the operation $*$ defined in Question 1d. Does every person have an inverse under this operation? Explain your reasoning.

Transforming Functions

You first learned about individual functions and then about families of functions. Next, you did arithmetic with functions, and then you composed them.

Now you will explore yet another way to work with functions: by transforming them. You will examine how transforming a function affects its various representations, in terms of symbols, tables, graphs, and real-world situations.

Candice Davis prepares a poster to display the effects of having transformed a familiar function.

Double Dose of Functions

You've looked at ways to combine functions to get new functions, using arithmetic operations and composition. You've also examined ways to use the graphs of individual functions to get the graphs of such combinations.

Now, rather than combining two or more functions at a time, you will look at ways to take a single function and alter it slightly. The functions created by these changes are called **transformations** of the original function. (There are other ways to transform functions besides those described here.)

In this activity, you will examine how these alterations affect the graph. Knowing how transformations affect functions and graphs will help you take familiar graphs and adjust them to fit specific situations. Although these questions ask about the graphs of functions, you may find it helpful to think about tables, equations, and real-world situations as well.

In Questions 1 and 2, you will begin with specific examples and look for a general principle.

1. If f is any function, we can define a related function g by doubling the output for f. In other words, we define g by the equation $g(x) = 2f(x)$. Your task is to examine how the graph of g is related to the graph of f. In each of parts a to c, graph the pair of functions on a single set of axes. You may find it especially helpful to use tables as an aid to understanding what's happening.

   a. Graph the functions defined by the equations $y = \sin x$ and $y = 2 \sin x$.

   b. Graph the functions defined by the equations $y = x^2 + 2x - 7$ and $y = 2(x^2 + 2x - 7)$.

   c. Pick another specific function for f and graph the functions defined by the equations $y = f(x)$ and $y = 2f(x)$.

   d. Based on your work in parts a to c, describe the relationship between the graph of a function and the graph of the function with twice the output. Then explain why this relationship holds true.

continued

2. For a given function f, we can define a related function h by doubling the input for f. In other words, we define h by the equation $h(x) = f(2x)$. Examine how the graph of h is related to the graph of f. Graph each pair of functions on a single set of axes.

a. Graph the functions defined by the equations $y = \sin x$ and $y = \sin (2x)$.

b. Graph the functions defined by the equations $y = x^2 + 3x - 7$ and $y = (2x)^2 + 3(2x) - 7$.

c. Pick another specific function for f, and graph the functions defined by the equations $y = f(x)$ and $y = f(2x)$.

d. Based on your work in parts a to c, describe the relationship between the graph of a function and the graph of the function with twice the input. Then explain why this relationship holds true.

Slide That Function

In *Double Dose of Functions,* you explored the effect on a graph of doubling either the output or the input of a function. Now you will investigate a different type of change to outputs and inputs.

Examine the effects of these two types of transformations:

- *Adding a fixed value to the output:* That is, if b is some fixed number, how does the graph of the function $y = f(x) + b$ compare to the graph of the function $y = f(x)$?

- *Adding a fixed value to the input:* That is, if c is some fixed number, how does the graph of the function $y = f(x + c)$ compare to the graph of the function $y = f(x)$?

Examine many examples, and be sure to include nonlinear functions. (For a linear function, it's hard to distinguish whether you have moved the graph vertically or horizontally.) Also be sure to consider both positive and negative values for b and c.

Write down the equations for each example you consider. Include rough but reasonably accurate sketches of their graphs. Summarize your results in a clear way, and explain why your conclusions are correct.

Transforming Graphs, Tables, and Situations

The activity *Double Dose of Functions* focused on the connection between the symbolic representation of a transformation and the effect of that transformation on the graph.

You will now examine another example involving graphs. But here, you will also look at how transformations are related to tables and the real-world contexts for functions.

1. The diagram shows the graph of the function defined by some equation $y = f(x)$.

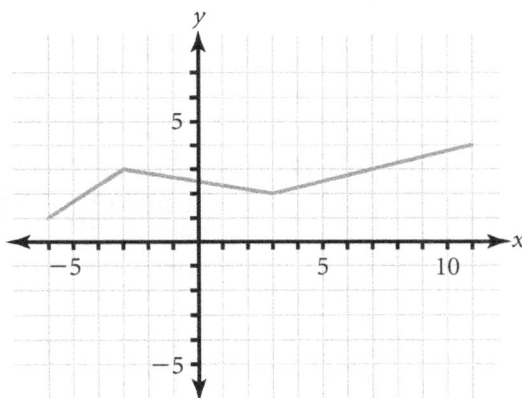

Trace four copies of the graph. Then draw the graph of each equation in parts a to d, putting each new graph on the same axes as one of the copies of the original graph. Note that some of these functions have a different domain from the original function.

a. $y = f(x) - 3$

b. $y = f(x + 2)$

c. $y = \frac{1}{2}f(x)$

d. $y = f(2x)$

continued ◆

2. The table shown here goes with the function defined by the equation $y = g(x)$.

x	$g(x)$
-2	7
-1	3
0	2
1	0
2	-3
3	-2
4	0
5	1
6	4

Use this table to create In-Out tables for each equation in parts a to d. Your tables may have different inputs from the original table, but make them as complete as possible based on the information you have about function g.

a. $y = g(x) + 2$

b. $y = g(x - 3)$

c. $y = 3g(x)$

d. $y = g\left(\frac{1}{2}x\right)$

3. Suppose the equation $y = h(x)$ gives the height in feet of a bouncing ball x seconds after it is released. Parts a and b ask you how the function would have to be transformed if units different from feet or seconds are used.

a. Let $k(x)$ represent the height of the ball, in inches, x seconds after the ball is released. Write an equation connecting $k(x)$ to $h(x)$.

b. Let $j(x)$ be the height of the ball, in feet, x minutes after the ball is released. Write an equation connecting $j(x)$ to $h(x)$.

Back to the Beginning

In the final activity of the unit, you will return to the opening problem and get the chance to apply some of the things you've learned. Well, it's not exactly the same as the opening problem. Instead, it's a more complicated—and more relevant—version of that problem.

Juan Madriz applies what he has learned about functions to the final problem of the unit.

Better Braking

This final activity of the unit revisits the unit's opening problem. In that problem, *Brake!,* you looked at a data set showing how far a particular car goes from the time the driver hits the brakes until the car comes to a complete stop. This stopping distance depends on the car's speed. But what really matters for safety is a little different.

Often, a driver needs to stop because she or he sees some danger ahead. But there is a delay between when a driver sees a danger and when the driver steps on the brakes. This delay is called *reaction time.* What really matters for safety is how far the car goes from when the driver sees the danger until the car stops.

This table gives that modified stopping distance.

Speed (in miles per hour)	Stopping distance (to the nearest tenth of a foot) *including distance driven during reaction time*
20	44.2
25	62.2
30	83.0
35	106.5
40	132.8
45	161.9
50	193.8

1. Explain what function family you think the data in this table represent, and why. Your explanation should involve both the table and the situation.

2. Find an algebraic equation that fits the table as well as possible. Be sure to test your equation on all the data items in the table.

Beginning Portfolio Selection

In your travels through *The World of Functions,* you looked at functions in these four ways:

- Using tables of values
- Using graphs
- Using equations
- In terms of real-world situations

Because these four representations are simply different ways of viewing the same thing, these perspectives have strong connections to one another.

Pick three activities from *The World of Functions,* each of which helped you see a connection between two of these ways of viewing functions. Each of the three activities should look at the connection between a different pair of perspectives. For each of the three activities, do each of these things:

- State which two perspectives the activity helped you connect.
- Describe how the activity helped you make the connection.
- Describe in your own words what that connection is.
- Describe some other activity you had in a mathematics class, this year or before, that made that same connection.

The World of Functions Portfolio

You will now put together your portfolio for *The World of Functions.*
This process has three steps:

- Writing a cover letter summarizing the unit
- Choosing papers to include from your work in this unit
- Discussing your personal mathematical growth in the unit

Cover Letter

For your cover letter for *The World of Functions,* focus on these two
key ideas:

- The distinguishing characteristics of each family of functions you
  worked with, including real-world situations, graphs, tables, and
  symbolic representations
- Methods of combining functions or transforming functions to create
  new ones

For each of these important aspects of the unit, choose an activity that
illustrates that idea to include in your portfolio.

Selecting Papers

Your portfolio from *The World of Functions* should contain these items:

- *Brake!, "Brake!" Revisited,* and *Better Braking*

  Include a statement of how your understanding of the "braking"
  situation developed over the course of the unit.

- *Linear Tables, Quadratic Tables,* and *Exponential Tables*

  Discuss how your understanding of the relationship between tables
  and algebraic representations grew through these three activities.

- Activities discussed in your cover letter

- *Beginning Portfolio Selection*

  Include the activities that you selected as part of this activity, along
  with your written work about those activities.

continued ▶

- A Problem of the Week

  Select one of the three POWs you completed during this unit (*One Mile at a Time, A Spin on Transitivity,* or *It's Off to College We Go*).

Personal Growth

Your cover letter for *The World of Functions* describes how the mathematical ideas develop in the unit. In addition, write about your own personal development during this unit. You may want to specifically address this question:

  How do you feel you have developed during this unit in terms of your ability to explore problems and prove conjectures in mathematics?

Include any other thoughts about your experiences that you wish to share with a reader of your portfolio.

SUPPLEMENTAL ACTIVITIES

The supplemental activities for *The World of Functions* continue the exploration of properties of functions and how to use combinations of functions. Here are some examples:

- *Absolutely Functions* and *Graphing Power* give you an opportunity to work further with the families of absolute value functions and power functions.

- *Odd or Even?* looks at two algebraic properties of functions and asks how those properties are related to the graphs of functions.

- *Ferris Wheel on a Ramp* and *"Small World" Again!* involve combining functions to fit real-world situations.

From Second Differences to Quadratics

In *Quadratic Tables,* you saw that if you make an In-Out table for a quadratic function using equally spaced inputs, the second differences of the outputs are constant. In this activity, you will explore the converse of that statement.

Suppose that *f* is a function. Suppose also that in any In-Out table for *f* using consecutive integers as the inputs, the second differences of the outputs are constant. Prove that for integers, the function is quadratic.

To begin, let *a* represent $f(0)$, let *b* represent the difference $f(1) - f(0)$, and let *h* represent the constant second difference. For instance, this second difference is the difference between $f(2) - f(1)$ and $f(1) - f(0)$. Use these symbols to find expressions for $f(2), f(3), f(4)$, and so on, and then look for a generalization.

Real Domains

Many different algebraic expressions can be used to define functions. For example, the expression $2x + 3$ defines the function f given by the equation $f(x) = 2x + 3$.

For some expressions, you can substitute any number for the variable. For instance, the expression $2x + 3$ makes sense no matter what value you substitute for x.

For other expressions, however, you need to be more careful. For example, consider the expression $\frac{1}{x^2 - 25}$. If x is either 5 or -5, the denominator of this fraction is 0, so the fraction is not defined. Therefore, the numbers 5 and -5 cannot be used in place of the independent variable for the function g defined by the equation $g(x) = \frac{1}{x^2 - 25}$. But any real number other than 5 or -5 can be used. (*Reminder:* The real numbers are the numbers represented on the number line. These include negative numbers, fractions, and irrational numbers, but not complex numbers.)

The set of number values for which an expression makes sense is the *domain* of the expression. That set is also the domain of the function defined by the expression. Thus, the domain of the function $g(x) = \frac{1}{x^2 - 25}$ is the set of all real numbers except 5 and -5.

Denominators are one source of limitations on the domain of a function. Square roots are another. For example, the expression $\sqrt{x - 3}$ makes sense only if $x \geq 3$. (If you consider complex numbers, this expression has a larger domain. For this activity, consider only real numbers.)

Part I

Find the domain for each expression. That is, find the set of real numbers for which each expression makes sense.

1. $\dfrac{x^2}{(x^2 + 1)(x + 3)}$

2. $\sqrt{x^2 - 16}$

continued

3. $\sqrt{x^3 + 8}$

4. $\dfrac{1}{\sqrt{2x - 5}}$

5. $\dfrac{x^2 - 9}{x\sqrt{x^2 - 3x + 2}}$

Part II

This time, you will be given a domain, and your task is to create an expression whose domain is the given set.

6. The domain is the set of all real numbers less than or equal to 6.

7. The domain is the set of all real numbers between -2 and 2 (including both -2 and 2).

8. The domain is the set of all real numbers between 1 and 4 (including both 1 and 4).

Absolutely Functions

You have come across the concept of absolute value in a variety of contexts. Intuitively, the absolute value of a number is the "purely numeric" part of the number, without regard to sign.

The absolute value of a number x is represented by the notation $|x|$. For example, $|4| = 4$, $|-7| = 7$, and $|0| = 0$.

Notice that $|x|$ is sometimes the same as x and sometimes the opposite of x. Formally, absolute value is defined as

$$|x| = \begin{cases} x \text{ if } x > 0 \\ 0 \text{ if } x = 0 \\ -x \text{ if } x < 0 \end{cases}$$

Note that in the last part of this definition, $-x$ is not a negative number. It represents the opposite of a number that is already negative.

Your main task in this activity is to investigate how absolute value affects functions and their graphs.

1. Suppose the table shown here comes from some function whose equation is $y = f(x)$. Based on this table, make a table for the function defined by the equation $y = |f(x)|$. Make your table as complete as possible.

x	$f(x)$
3	7
−2	−9
−6	11
8	0
0	−12
1	−5
7	2
−5	8
11	7

continued ▶

2. Sketch the graph for each function. Show the scales on both axes. Keep in mind that these functions may not be defined for all real numbers.

a. $g(x) = |\sin x|$

b. $w(d) = \left|\dfrac{1}{d}\right|$

c. $z(r) = |\log r|$

d. $h(s) = \log|s|$

3. Suppose the graph represents the function defined by the equation $y = k(x)$. Sketch the graph of the function defined by the equation $y = |k(x)|$.

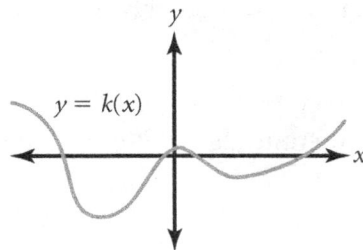

4. Now consider again the function f whose table is given in Question 1.

a. Graph the points that are given for f in that table.

b. Make an In-Out table for the function defined by the equation $y = f(|x|)$. Make your table as complete as possible with the information you have.

c. Draw a new set of axes identical to those you used in Question 4a. Graph the function $y = f(|x|)$ on this set of axes.

5. Consider the function k from Question 3. Make a graph for the function defined by the equation $y = k(|x|)$.

Odd or Even?

You know what *odd* and *even* mean when you're talking about integers. For example, 4, 26, 0, and −10 are even integers, while 17, −1, 95, and −513 are odd integers.

The same words are used to define certain types of functions.

- A function *f* is an *even function* if it has the property that $f(x) = f(-x)$ for all values of *x* in the domain of *f*.
- A function *f* is an *odd function* if it has the property that $f(-x) = -f(x)$ for all values of *x* in the domain of *f*.

These definitions are rather abstract, so here's an example of each type.

The function *g* defined by the equation $g(x) = x^2$ is an even function, because $g(x) = g(-x)$. For instance, $g(5) = g(-5)$, because both $g(5)$ and $g(-5)$ are equal to 25.

The function *h* defined by the equation $h(x) = \sin(x)$ is an odd function, because $h(-x) = -h(x)$. For instance, $h(-30°) = -h(30°)$, because $\sin(-30°) = -0.5$ and $\sin(30°) = 0.5$.

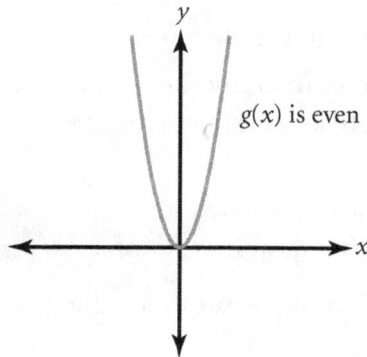

$g(x)$ is even

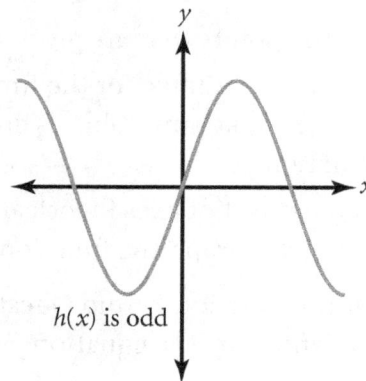

$h(x)$ is odd

continued ▶

1. Find at least five more examples of even functions and at least five more examples of odd functions. Vary your examples for each category as much as possible.

2. Sketch the graphs of at least three functions in each category.

3. a. What general properties does the graph of an even function have?

   b. What general properties does the graph of an odd function have?

4. Give an example of a function that is neither even nor odd.

5. Think about the special case of functions of the form $k(x) = x^n$, for different whole-number values of n. What insights do these functions give you into the reason for the terms *odd function* and *even function*?

Graphing Power

In this activity, you will explore the graphs for functions of the form $f(x) = ax^b$, where a and b are constants. These functions are called **power functions.**

The functions g and h defined here are both power functions.

$$g(x) = x^3 \qquad h(x) = -2x^{0.5}$$

In some contexts, it makes sense to use the term *power function* only when the exponent b is a whole number. For this activity, you should consider all values.

Begin by looking only at power functions in which the coefficient a is equal to 1 and the exponent b is positive. Focus your attention especially on the part of the graph between $x = 0$ and $x = 1$.

1. How do the graphs of power functions of the form $y = x^b$ behave between $x = 0$ and $x = 1$ for different positive values of b?

2. How do these graphs behave outside the set of values between $x = 0$ and $x = 1$?

3. How would the graph of a power function change if the coefficient a were different from 1?

Ferris Wheel on a Ramp

In the unit *The Diver Returns,* you explored height on a Ferris wheel for various radii and periods. But now there's a new complication.

The Ferris wheel itself has been placed on a ramp, as shown here. The entire Ferris wheel gradually slides down the ramp as the Ferris wheel goes around. The ramp is 27 feet high at its upper end, and the base of the ramp extends 135 feet along the ground.

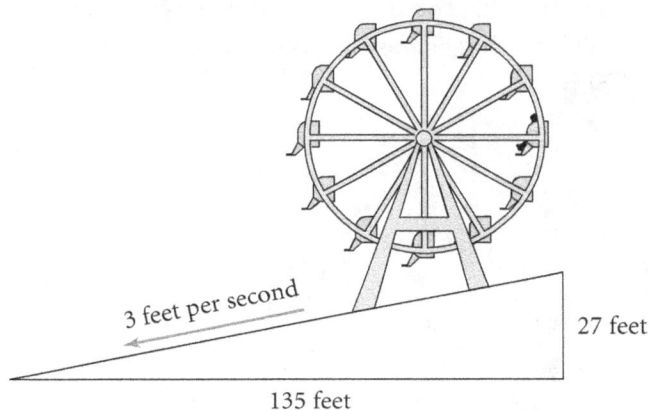

When the Ferris wheel starts sliding, its center is 45 feet above the ground. The entire Ferris wheel moves down the ramp so that its center travels 3 feet per second, in a direction parallel to the slant of the ramp itself. The Ferris wheel has a radius of 20 feet and turns counterclockwise with a period of 12 seconds.

Suppose a rider is at the Ferris wheel's 3 o'clock position when the wheel starts sliding. Find an expression for the rider's height off the ground after t seconds. Graph the function defined by this expression from $t = 0$ to $t = 24$.

Freddie on the Ferris Wheel

Al owns a miniature Ferris wheel, which runs on a battery. One day he brings his miniature Ferris wheel to the amusement park, along with his teddy bear, Freddie.

Of course, Freddie likes to ride in his own little seat, so Al puts Freddie in one of the seats of the miniature Ferris wheel and then gets on the big Ferris wheel.

Throughout the ride, Al holds the miniature Ferris wheel so that its base is parallel to the big wheel's base and so that its center is at the same height as Al himself.

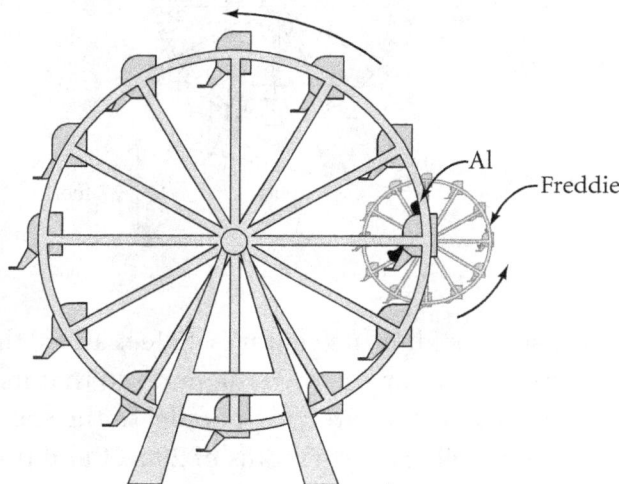

The two Ferris wheels start turning simultaneously. Al is in his usual place, level with the center of the Ferris wheel. Freddie is also level with the center, as shown in the diagram. Both Ferris wheels turn counterclockwise.

The big Ferris wheel has a radius of 30 feet and a period of 20 seconds. The small Ferris wheel has a radius of 10 feet and a period of 10 seconds.

Find an expression for Freddie's height, relative to the center of the big Ferris wheel, after t seconds. Then graph the function defined by this expression from $t = 0$ to $t = 40$.

Over, and Over, and Over, and . . .

Functions can be composed with themselves as well as with other functions. For example, if f is defined by the equation $f(x) = 3x + 2$, we can form functions such as $f \circ f$ and even $f \circ f \circ f$.

For instance, $(f \circ f)(x)$ means $f(f(x))$, so $(f \circ f)(x) = f(3x + 2)$. This equals $3 \cdot (3x + 2) + 2$, which simplifies to $9x + 8$.

1. Verify that $(f \circ f \circ f)(x) = 27x + 26$.

We generally abbreviate $f \circ f$ as f^2, $f \circ f \circ f$ as f^3, and so on. This notation can be misleading. For instance, f^2 could mean $f \cdot f$ rather than $f \circ f$, but the notation f^2 is generally used specifically for composition.

The process of composing a function with itself is called *iteration* or *iterating the function.*

Interesting things can happen with iteration. One thing to investigate is what happens if you start with a particular value for x—say, $x = c$—and then find $f(c), f^2(c), f^3(c)$, and so on.

If $f(c) = c$, we say that f has a *fixed point* at $x = c$ (or that c is a fixed point for f). For example, the function f defined by the equation $f(x) = 3x + 2$ has a fixed point at $x = -1$, because $f(-1) = -1$. For this function, if $c > -1, f^n(c)$ grows toward infinity as n increases, while if $c < -1, f^n(c)$ grows toward negative infinity as n increases.

2. Suppose c is a fixed point for a function g. Explain why this guarantees that $g^n(c) = c$ for all positive integers n.

3. Suppose f is defined by the equation $f(x) = 3x + 2$. Find the numeric value of each of these expressions.

   a. $f(2), f^2(2), f^3(2)$, and $f^4(2)$

   b. $f(-3), f^2(-3), f^3(-3)$, and $f(-3)$

continued ▶

4. Consider the function g defined by the equation $g(x) = x^2$.

   a. Find out whether g has any fixed points.

   b. Describe what happens to $g^n(c)$ as n increases, for values of c other than the fixed points. (What happens may be different for different values of c.)

5. Choose two other functions. For each of your functions, do two things.

   a. Find the fixed points of your function (if any).

   b. Describe what happens for various values of x, other than fixed points, as you keep iterating the function.

Its Own Inverse

1. In *Linear Functions in Verse, f* was an arbitrary linear function defined by the equation $f(x) = ax + b$ (with $a \neq 0$). You needed to find an expression for $f^{-1}(x)$.

   Which linear functions are their own inverses? That is, for what choices of a and b is f equal to f^{-1}? (*Note:* There are infinitely many possibilities.)

2. In *Functions in Verse*, you were asked to describe the relationship between the graph of a function and the graph of its inverse.

   a. How could that relationship be used to help with Question 1 of this activity?

   b. Can you find any nonlinear functions that are equal to their own inverses?

A Hyperbolic Approach

In Question 1 of *Story Sketches III,* you considered this problem:

> You are driving due north on the highway. Ahead, but off to the east, you see a tall tower. Sketch a graph and make a table showing your distance from the tower as a function of time as you continue north.

In this activity, you will get more precise about the situation.

Suppose the tower is 100 feet east of the road and you are going 60 feet per second (roughly 40 miles per hour). At time $t = 0$, you are 1000 feet south of the point on the road that is due west of the tower.

1. Develop an equation that gives your distance d from the tower as a function of t. Express d in feet and t in seconds.

2. Equations of the form $ay^2 - bx^2 = 1$, where a and b are positive numbers, give graphs that have a shape called a *hyperbola*. For example, the equation $y^2 - x^2 = 1$ has the graph shown here. Each of the two parts of the graph is called a *branch* of the hyperbola.

   Use ideas about transformations of functions to explain why the graph of your equation in Question 1 is a branch of a hyperbola.

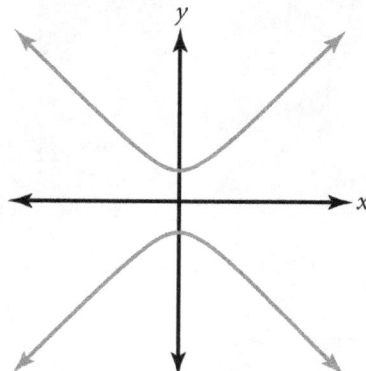

Small World Again!

Now that you've spent all this time learning about functions and how to put them together, you get one more opportunity to examine the population data set from *Small World, Isn't It?* Here is the data set.

Year	Estimated population
1650	470,000,000
1750	629,000,000
1850	1,030,000,000
1900	1,550,000,000
1950	2,560,000,000
1960	3,040,000,000
1970	3,710,000,000
1980	4,450,000,000
1990	5,280,000,000
2000	6,090,000,000
2005	6,480,000,000

Source: U.S. Census Bureau.

Use whatever techniques you can think of to find a function that fits this data set as well as possible, based on the **least-squares method** of evaluating approximations.

Write a report describing your work, your reasoning, your conclusions, and your final function.

PHOTOGRAPHIC CREDITS

Front Cover Photography

(upper and middle rows) Stephen Loewinsohn; (lower row) iStockphoto

The World of Functions

1 (upper and middle rows) Stephen Loewinsohn; (lower row) iStockphoto; **3** Stephen Loewinsohn; **4** iStockphoto; **5** Shutterstock; **6** iStockphoto; **7** Peter Arnold, Inc./Alamy; **9** Shutterstock; **10** Stephen Loewinsohn; **14** Shutterstock; **15** iStockphoto; **25** iStockphoto; **26** iStockphoto; **27** Stephen Loewinsohn; **28** Stephen Loewinsohn; **36** Stephen Loewinsohn; **37** (upper) iStockphoto; **37** (lower) iStockphoto; **43** Sandie Gilliam, Lynne Alper; **47** iStockphoto; **49** Stephen Loewinsohn; **51** Stephen Loewinsohn; **56** Photodisc; **57** Stephen Loewinsohn; **61** Shutterstock; **73** Sandie Gilliam, Lynne Alper; **76** Stephen Loewinsohn; **79** Stephen Loewinsohn; **87** iStockphoto; **96** Shutterstock; **97** Shutterstock; **99** Photodisc

www.ingramcontent.com/pod-product-compliance
Lightning Source LLC
Chambersburg PA
CBHW051226200326
41519CB00025B/7261